Plants and How They Grow

SCHOOL PUBLISHERS

Orlando Austin New York San Diego Toronto London

Visit *The Learning Site!*
www.harcourtschool.com

Two Types of Plants

Look at this photograph of the rain forest in the Amazon River basin in Brazil. Picture yourself walking beneath the towering trees. How many different types of trees and other plants do you think you would see? A part of a rain forest no larger than a city block may have more than 400 different types of trees! Most forests in the United States have no more than 20 different kinds of trees in the same area.

Whether they are in the forests of Brazil, the forests of the United States, or any other place in any other country, all plants can be divided into two main groups: vascular plants and nonvascular plants. Vascular plants, such as trees, have vascular tissue. **Vascular tissue** supports plants and carries water and food. Nonvascular plants, such as mosses, do not have vascular tissue to carry water and food.

Vascular plants grow in a variety of climates, from cacti in the desert, to orchids in this rain forest.

Nonvascular plants do not have true roots, stems, or leaves, although they do have structures that look like each of those parts. Nonvascular plants can only pass water, nutrients, and food from one cell to the next. For this reason, nonvascular plants are small. Vascular plants have true roots, stems, and leaves. These structures allow for a greater variety in the size and appearance of vascular plants.

MAIN IDEA AND DETAILS What are the two main types of plants?

When you see moss on a rock or a tree, you are seeing many tiny moss plants living close together.

Roots, Stems, and Leaves

There are more than half a million types of vascular plants on Earth. They range in size from tiny duckweed (less than $\frac{1}{4}$ inch long) to giant redwoods (taller than a 25-story building)! No matter how different they appear, all vascular plants have three parts in common—roots, stems, and leaves.

Roots absorb water and nutrients from the soil and are a plant's anchor. Different types of roots are adapted to different environments. Some plants have taproots—one large, strong root that grows deep into the soil. Some taproots are used for the plant's food storage. Other plants have fibrous roots, which are thin and branching and form a mat below the ground's surface. They absorb water from a large area. Fibrous roots help prevent soil erosion by wind and water by holding soil in place.

Fast Fact

Buffalo grass has a thick network of roots. These roots and the soil they hold are called sod. The roots hold soil so well that many settlers on the American plains used sod to build their homes.

Some taproots, such as carrots and beets, store so much nutrition that people use them for food!

Stems do several things for plants. They hold the plant up and support its leaves so that they will be in sunlight. Stems also carry water and nutrients between the roots and leaves of a plant.

Not all stems look alike. Small plants usually have soft, green, flexible stems. The water inside the stem makes it firm enough to hold the plant up. Trees and other tall plants have stiff, woody stems for extra support. Some desert plants store food and water in fleshy stems.

MAIN IDEA AND DETAILS
How does structure relate to function in different plants' roots and stems?

Large plants, like these redwood trees, get extra support from woody cells in their stems. Redwoods may live for hundreds of years.

Leaves and Light

Roots, stems, and leaves work together for a plant's growth and survival. You've already learned that roots and stems support a plant and transport water and nutrients. What do leaves do? They make a plant's food!

Leaves make food through a process called **photosynthesis**. In this process, plants use water and nutrients from the soil, carbon dioxide from the air, and energy from sunlight to make sugar. Photosynthesis also produces oxygen, which plants release into the air.

The outer layer of cells in a leaf is the epidermis. It protects the leaf from damage.

Photosynthesis takes place inside chloroplasts in certain leaf cells. A chloroplast contains a green material called chlorophyll that absorbs sunlight. In addition to chloroplasts, leaves contain veins, or small bundles of vascular tissue. Veins bring water and nutrients to the chloroplasts. They also carry the sugar produced by photosynthesis throughout the plant. Carbon dioxide, which the plant needs for photosynthesis, enters a leaf through tiny holes called stomata. Stomata open and close to let carbon dioxide in and oxygen out. Stomata are found on the underside of most leaves.

MAIN IDEA AND DETAILS **What is the primary role of a plant's leaves?**

Leaves come in many shapes and sizes. Most leaves are flat and thin, which helps them absorb sunlight.

Mosses and Ferns

You have learned that there are many different types of plants, and they grow and survive in different ways. Different plants not only grow in different ways but they also reproduce in different ways. Most plants reproduce by means of spores or seeds.

Fast Fact

Did you know that about 325 million years ago, vast forests of tall tree ferns covered much of Earth? Now most ferns are found in the tropics.

Mosses and ferns are two types of plants that reproduce by spores. A **spore** is a single reproductive cell that can grow into a new plant. Remember that mosses are simple, nonvascular plants. Ferns are vascular plants.

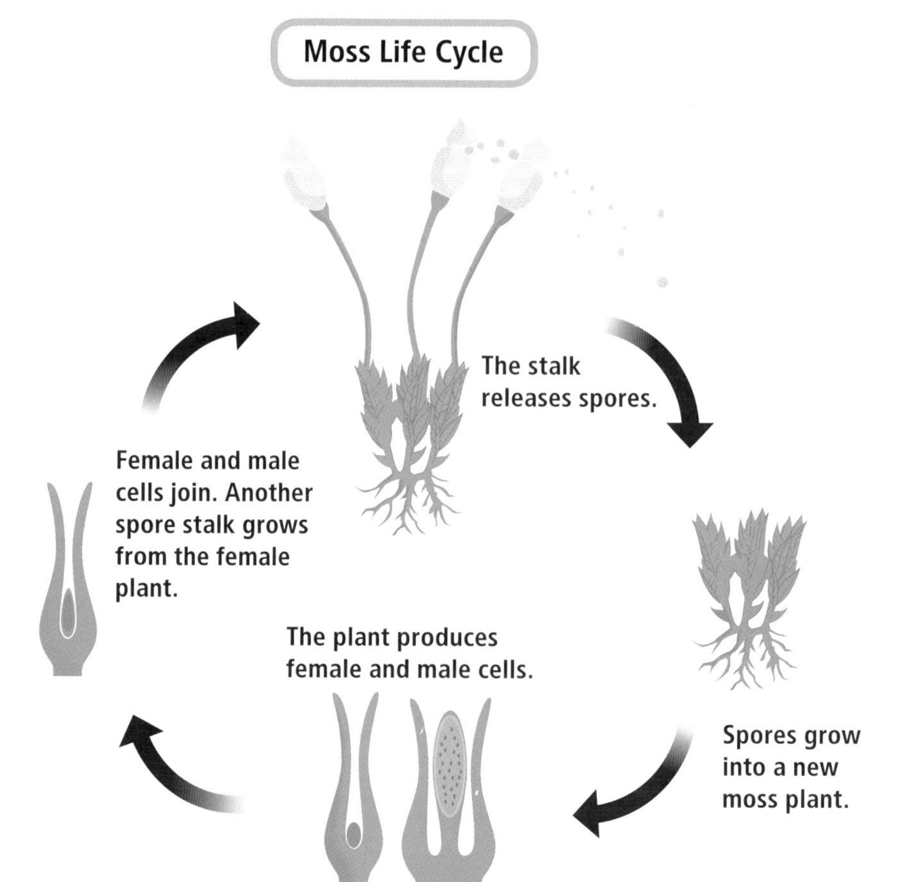

The reproductive cycles of mosses and ferns are similar. They both release spores that grow into tiny plants called gametophytes. These have female and male reproductive structures. In both types of plants, female and male cells join together to grow into a sporophyte, which makes more spores.

Although the reproductive cycles of mosses and ferns are similar, they are not the same. In mosses, male and female reproductive cells are produced on separate plants. A female and male cell unite and produce a stalk that grows out of the female plant. The stalk releases the spores that will grow into new moss plants. In ferns, each gametophyte has both male and female reproductive structures. A male reproductive cell swims to another gametophyte to join with a female reproductive cell. The united cell then divides and grows into a spore-producing plant.

COMPARE AND CONTRAST **How are mosses and ferns alike? How are they different?**

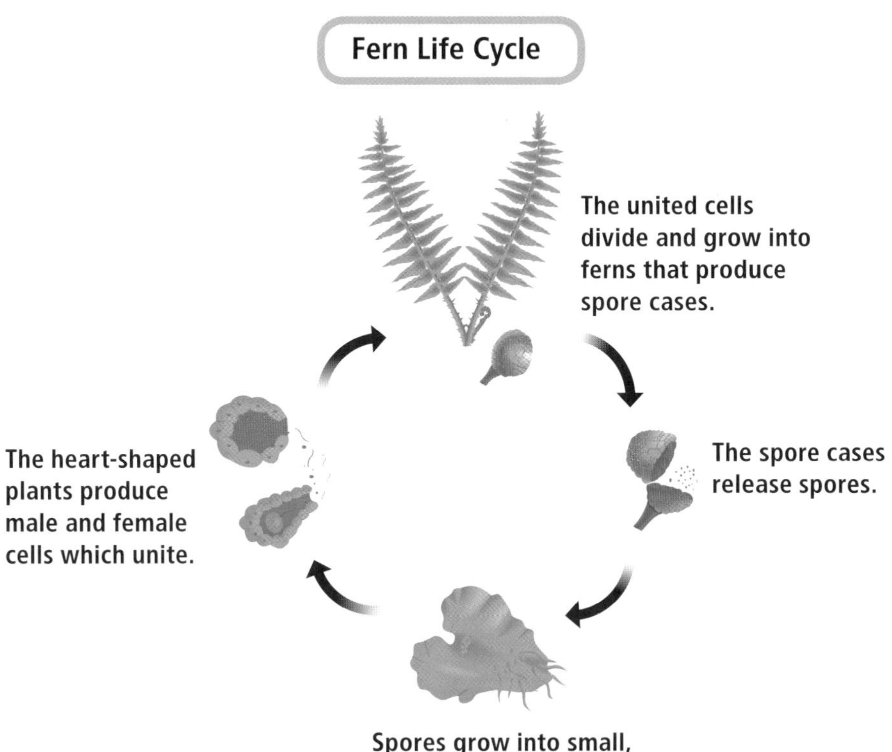

Plants with Seeds

While simple vascular plants, such as ferns, reproduce with spores, most vascular plants reproduce by making seeds. Plants that make seeds are classified into two groups. Flowering plants produce seeds in flowers. Some flowers produce seeds inside fruit. Flowering plants that have seeds protected by fruit are classified as **angiosperms**. Other plants produce seeds that are considered "naked," because they are not protected inside a fruit. Plants that produce naked seeds are classified as **gymnosperms**.

A fruit protects seeds in several ways. It helps keep birds and other animals from getting them and serves as a covering that protects seeds from cold weather.

The most common gymnosperms are the conifers, or cone-bearing plants. A conifer that may be familiar to you is the pine tree. Have you ever seen a large pine tree with cones hanging from its branches? Conifers produce seeds in the cones.

Most conifers produce both male and female cones on the same tree. Female cones are larger and grow high on the trees, above the male cones. Wind carries pollen from the male cones to the female cones. The pollen contains male reproductive cells. The male and female reproductive cells unite and develop into seeds. When the seeds are mature, the cone scales open and the seeds are released. If a seed lands in a suitable habitat, a new tree grows, and the life cycle begins again.

COMPARE AND CONTRAST How are the seeds of gymnosperms and angiosperms different?

Seeds develop between the scales of the cones. Conifers are common in several climates.

All About Flowers

Seed development is not as simple in angiosperms as in gymnosperms. On a gymnosperm such as a pine tree, separate cones on the plant produce the female and male reproductive cells. But in many flowering plants, the male and female reproductive organs are often together in the same flower. Unlike gymnosperms, which are pollinated only by the wind, angiosperms can also be pollinated by insects and other small animals. The color, shape, and scent of flowers attract animals to them. Flowers also produce a sugary nectar that attracts and feeds many different insects and birds. These characteristics of flowers attract animals, making it more likely that the animals will transfer pollen.

Fast Fact

Did you know that there are more than 235,000 kinds of flowering plants? These include grasses, herbs, shrubs, and many trees. Flowering plants are important sources of wood, fiber, and medicine. Much of the food that people eat comes from flowering plants!

This tree is an angiosperm. It produces its seeds in flowers instead of cones.

This picture identifies some parts of a flower. Anthers make pollen, which sticks to insects or floats on the wind. Stigmas collect pollen from another flower.

How do insects or other animals pollinate flowering plants? Picture a bee buzzing around a field of flowers. Bees feed on the nectar from flowers. As a bee crawls into a flower to get nectar, pollen sticks to the bee's hairy body. The pollen is from some of the flower's male parts, called anthers. When a bee goes into another flower, some of the pollen on its body rubs off on a female flower part called the stigma. If the pollen from the male flower part successfully gets to the female flower part, fertilization may occur. The male and female reproductive cells unite. The flower then develops into a fruit that contains seeds.

COMPARE AND CONTRAST What are the differences between seed development in flowers and seed development in cones?

How a Seed Grows

A seed can survive inside its thick, waxy seed coat for several years until the conditions are right for it to grow. These conditions usually include fertile soil, warm temperatures, and moisture. When conditions are right, a seed will sprout, or **germinate**.

First, a seed absorbs water and expands. As the seed swells, the seed coat splits. The tiny plant within the seed begins to grow. The first part to develop is the root, which anchors the plant and begins to take up water. Next, the stem begins to grow upward, toward the light. The first leaves, the seed leaves, are attached to the stem. At this stage the growing plant cannot make its own food. Instead, it uses food stored in the seed leaves.

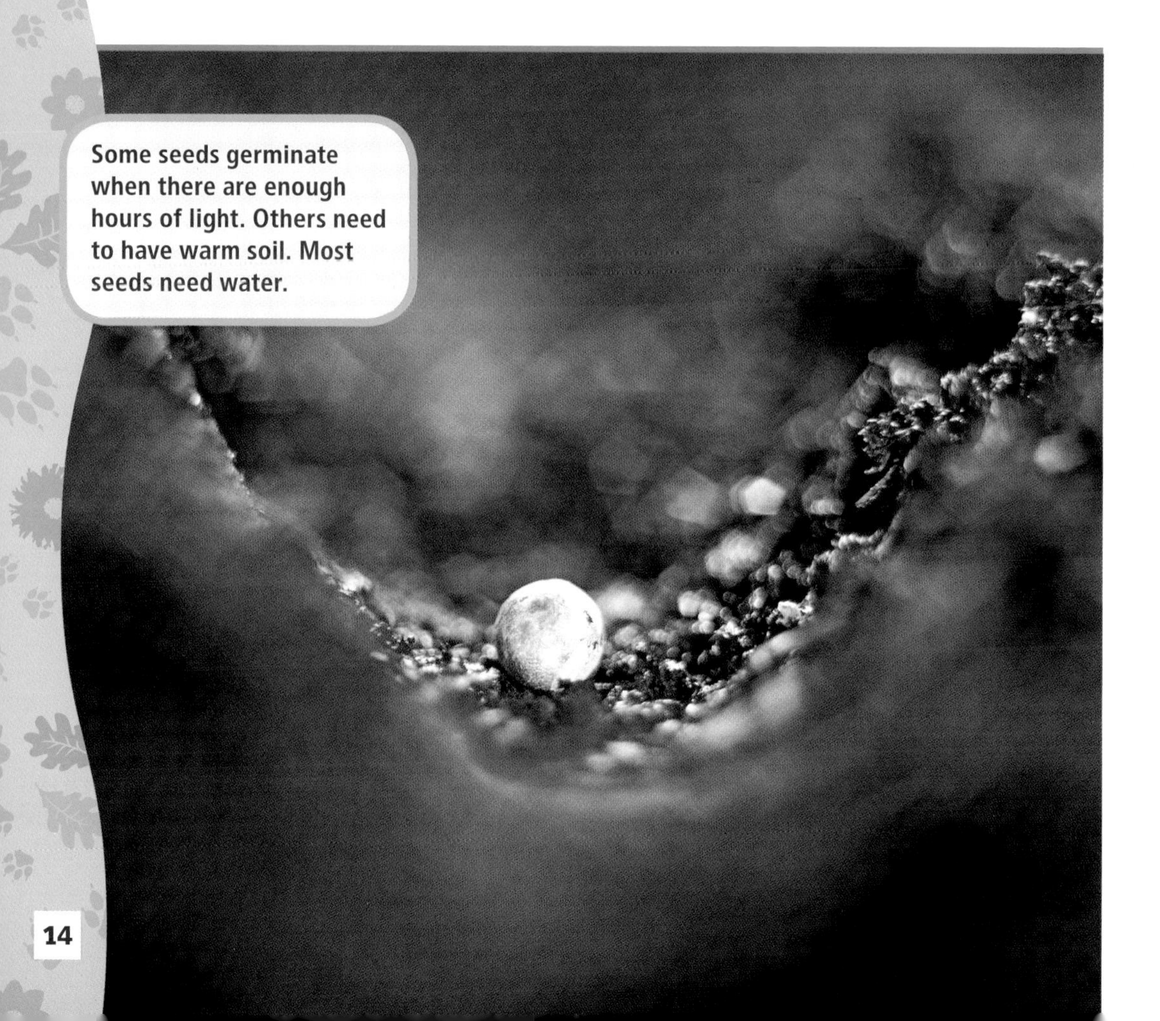

Some seeds germinate when there are enough hours of light. Others need to have warm soil. Most seeds need water.

As it grows, the plant produces longer and thicker roots. The stems get taller and stronger. The plant's true leaves emerge and begin producing food through photosynthesis. When the plant is growing well and its leaves are making all the food the plant needs, the seed leaves drop off. The plant can now live on its own.

COMPARE AND CONTRAST How does the young plant differ from the mature plant?

By the time the small plant emerges from the soil, it has a well-developed root system and its true leaves are ready to produce food.

Summary

All over the world, there are many different kinds of plants. Scientists classify plants into two main groups by the way they transport water. Vascular plants have vascular tissue that carries water from one part of the plant to another; nonvascular plants do not. Scientists also classify plants by the way they reproduce. Nonvascular plants and simple vascular plants reproduce by spores. Gymnosperms and angiosperms are vascular plants that reproduce by seeds.

Glossary

angiosperm (AN•jee•oh•sperm) A flowering plant that has seeds protected by fruits (10, 11, 12, 15)

germinate (JER•muh•nayt) To sprout (14)

gymnosperm (JIM•noh•sperm) A plant that produces naked seeds (10, 11, 12, 15)

photosynthesis (foh•toh•SIHN•thuh•sis) The process in which plants make food by using water and nutrients from the soil, carbon dioxide from the air, and energy from sunlight (6, 7, 15)

spore (SPAWR) A single reproductive cell that can grow into a new plant (8, 9, 15)

vascular tissue (VAS•kyuh•ler TIH•SHOO) The tissue that supports plants and carries water and food (2, 7, 15)